The Dimension Bible

For The Remodeler and Do It Yourselfer

**John A. Knoelk
aka "Contractor John"**

PRESS

The Dimension Bible
For The Remodeler and Do It Yourselfer
by John A. Knoelk aka "Contractor John"

Printed in the United States of America
Edited by Xulon Press

ISBN 9781498412193

www.xulonpress.com

Table of Contents

Welcome to *The Dimension Bible*, the complete handy reference guide for all your remodeling and building projects. This book is the result of my forty years of remodeling and building experience. While no book can address every measurement for every situation, I believe within these pages you will find the vast majority of measurements, dimensions and formulas to calculate and arrive at the correct answer you will need to complete your project. In most instances I have included the actual measurement or dimension; in others I have included a formula so you can plug in your own numbers and arrive at the answer.

You will find the guide to be user friendly. I have divided this reference guide into four sections. The first section contains general information useful in designing a complete home or a particular space within it. The second section breaks down each room and supplies you with the dimensions for items in each of those rooms. The third section contains charts, tables, and formulas so you may arrive at a result that applies to your unique situation. The fourth section talks about items on the exterior of the house. One of the problems I usually have with this type of book is that there isn't enough space for notes. I have purposely left each left hand page empty so there is plenty of room for you to add your own notes.

When utilizing the charts and formulas, make sure you are using the correct room information for the particular room you are upgrading. For example, you shouldn't use the kitchen exhaust fan chart to calculate the size of the bathroom exhaust fan.

I have purposely refrained from adding a lot of written word, and have strived to be as clear and concise presenting the sometimes-confusing information. As with any reference book it's an ongoing, evolving project. If you have anything to add to the book or if you find a special circumstance that I have not addressed, please send me the information at dbsuggestions@dimensionbible.com and I will do my best to include it in future releases.

Contractor John

Dedication

I wish to dedicate this, my first book, to my Lord and Savior, Jesus Christ, who has been by my side continually, sometimes to the point of carrying me.

To my parents Erwin "Bud" and Yvonne Knoelk for providing a loving, stable home life and instilling in me my work ethic and my stick-to-ed-ness. To my dad "Bud" for his willingness to share in my dream of building a two-level tree house and then helping me to build it. I was hooked. Also for all the help and advice I have received along this journey called life. To my wife Sue, who has supported, encouraged and steadfastly stood behind me, sometimes pushing me, but always with love.

Legal Disclaimer

I was told I have to include this legal section, which basically says that to the best of my knowledge I believe all the information included in this book to be true and accurate. In the event you use this information and something goes awry, you can't sue me. The information contained within this book is partly my opinion drawn from my years of personal experience, the experience of other trades people I have known through the years, industry standards, national, state and local building codes, which may or may not be applicable in every locale or situation. It is completely your responsibility to check with your local building department before beginning any work. It is your responsibility to do the appropriate research on all aspects of your project before you begin.

Construction Abbreviations

Abbreviation	Meaning	Typical Uses
T&G	Tongue and Groove	Plywood/Flooring
O.C.	On Center	Wall Stud/Joist/Rafter Spacing
AFF	Above Finished Floor	Finished materials are set to this measurement
TOF	Top of Foundation	Exterior Grade is typically set 6" below TOF
GFA	Gas Forced Air	Type of furnace
A/C	Air Conditioning	
HVAC	Heating, Ventilation, Air Conditioning	
ADA	Americans with Disabilities Act	Sets minimums and requirements for accessibility
LV	Low Voltage	Phone/TV/Cable/Alarm/Sound etc. type wiring
RO	Rough Opening	The framed opening that a typical door or window would fit into
GFCI	Ground Fault Circuit Interrupter	Sometimes referred to as a GFI
PSF	Pounds per sq. ft.	To calculate loads
CFM	Cubic Feet per Minute	To measure air movement
"T"	A plumbing fitting that has 3 inlets and looks like a "T"	
GLA	Gross living area	Used to calculate legal occupiable space

Section 1

Designing Your Project

General Design Principals

Many books have been written and consequently there are many opinions on house and individual room design. I believe the most important factor in any design is the traffic pattern or flow—whether it's an entire house or one room.

Secondly, you should always design the space to the person or people that will be utilizing the room on a regular basis. An example would be: remodeling a bathroom where the individuals using it every day are tall; you would want to raise the cabinet and mirror height, etc. Remember that individual needs should be part of any project or plan.

Another equally important aspect is utilizing sight lines to create mystery or visual areas of interest. An example of this is when you enter a home you shouldn't be able to stand there and see everything, there should be things blocking your sight line, something that creates a visual mystery. If you turn the corner of a hallway and immediately view an open doorway and a toilet, you have failed design 101. Instead you should design a floor plan that will create a situation where you have to walk to the door and enter to see the full view of the bathroom—this creates interest of design.

Interesting lines on the exterior are just as important. The mix of hip and gable roof styles, small sections of wall bump outs and height differences give the feel of depth and tend to make a home look bigger, adding curb appeal.

Finally, always plan and design on paper before you begin to remodel or build. When the two-dimensional plan leaves paper and becomes reality, you may see facets of the design you have missed, so you will always have to be prepared to change the design. If you have taken all the necessary design steps and thought it through, the "field changes" will be kept to a minimum.

Universal Design

Universal design is a relatively new term and as such many homeowners don't really understand what it is or isn't. Simply stated, universal design isn't just for disabled people. It makes all of our lives easier, through the use of products and design, no matter what your age or physical condition. Universal design is a style of building or remodeling that can include: design specific ingredients such as aging in place, smart home technology, and space planning.

Let me illustrate using a few examples we all can relate to. Remember back when you had to carry your books to school under your arm? Along came backpacks, and allowed you to carry all your books, laptop, and even your lunch. Growing from that simple idea the front-pack baby carriers were developed, both of which are a form of universal design, once again making life easier for everyone. Another example of a universal design product we all can relate to is the small wheels that are installed on suitcases. These wheels applied to the bottom of our suitcases allow you to haul your luggage through an airport or hotel hallway with greater ease.

If we take a look at universal design in our homes we can find many products or features that a mere ten years ago were absent. Some of them may have been used commercially or in situations where you were building or remodeling for disabled people. Infrared sensors on faucets and hand dryers are examples, as are infrared sensors on toilets. Lever handles on faucets and doors have been used in doctors' offices and medical facilities for years, but are now finding their way into our homes.

Looking at the inroads smart technology has made into universal design is as easy as looking at your HVAC system. You can now turn your HVAC system on and off and even adjust the temperature from your smart phone or laptop. You can monitor your alarm system from your smart phone and receive text messages when an alarm is tripped. You can be notified when a predetermined event takes place such as low temperature or sump pump failure.

Certain types of lighting controls fall under universal design, such as dimmers. I am sure we all have walked into a bathroom and had the light go on magically. Obviously it was controlled by a wall switch that senses motion and turns on, and after a predetermined amount of time the sensor will turn the lights off, once again making life easier and additionally saving energy. Doorbell intercom systems that ring to the phones in the house are yet another example of universal design.

I'm sure you can think of examples in your own life and home that fall under the universal design umbrella, making our lives easier. If you think about it, isn't that the need the vast majority of inventions address, making our life easier and more comfortable. Incorporating universal design into all your remodeling and building projects will make your lifestyle easier and will add to the value of your home when and if you decide to sell it.

Kitchen Design

Have you ever lived in a home where the kitchen just "worked" and cooking there was effortless? Or conversely everything you tried to do in the kitchen didn't work and nothing was where it should be?

The answer is simple: "proper kitchen design principles." Kitchen design principles based on research in the 1920's still work in today's homes. They are still considered the "gold standard" when designing a kitchen.

Today's kitchens are used more than ever, for various activities and functions so it is imperative that adequate planning goes into their design. Obviously cooking still remains the main function of a kitchen, but areas for visiting, kids doing homework, paying bills, or using the computer are becoming increasingly common.

Several general pre-design questions to ask are: does traffic flow through the kitchen or does it start or stop there? How many people generally will be cooking in the space? The more questions you ask and answers you have, the better the function-ality of your new kitchen will ultimately be.

To begin your design, draw out the perimeter walls defining the space along with all adjacent rooms and their use to provide traffic pattern information. Make sure to include any island knee walls or peninsulas. Make sure to define the width of all openings to and from the kitchen. Door openings should be a minimum of a thirty-two inch clear opening. If the opening is a cased opening or a drywall opening, measure from wall to wall. If there is a door involved, open the door ninety degrees and mea-sure from the door face to the doorstop on the opposite side, remember you want thirty-two inches. If a cabinet runs parallel to the opening and "the hallway" created by the cabinet and adjacent wall exceeds twenty-four inches in depth (length), then the width of the doorway needs to expand to thirty-six inches. If a stove is the item creating "the hallway", make sure it is buffered from "the hallway" by a twelve inch cabinet and there is minimum of fifteen-inch-wide clear countertop space on the side opposite "the hallway" to place anything hot that you remove from the stove. One last point is you will want to make sure that any cabinet doors and drawers don't interfere with the operation of any appliance.

Aisles in the kitchen should be a minimum of forty-two inches wide for one cook and forty-eight inches wide if there are going to be two cooks working at the same time. Aisle widths should be measured from the leading edge or protruding handle of any item to a point directly across from it. Major through-traffic should not flow through the work aisle. If you find that this situation does exist, then redesign the area to eliminate that pattern

The Infamous Work Triangle

Have you ever heard of the work triangle? The kitchen work triangle is formed by drawing a line from the front center of the refrigerator, the front center of the stove, and the front center of the sink, connecting them to form a triangle. No single leg of the triangle should be less than four feet or more than nine feet and the total of all the legs of the triangle should not exceed twenty-six feet. You will want to avoid placing any two of these items alongside each other on the same wall. Multiple work zones and multiple work triangles can be established with the addition of dishwashers, trash compactors, etc.

After you have placed the appliances, you can begin to place the appropriate cabinets between the appliances. Take care to place the correct configuration of cabinets in each space. Drawers that hold silverware and utensils should be located close to the dishwasher, while cabinets for storing large pots and pans should be located by the stove.

Seating

We have all been in a kitchen with tight seating where somebody is trapped in the corner. The minimum space between the edge of the table and the wall should be thirty-two inches if there is no traffic behind the person. If there is traffic moving behind the chair, then a minimum of thirty-six inches is needed, and forty-four inches if you prefer to have a person pass by without turning sideways. Always remember to measure from the front edge of any protruding object.

People tend to gather in the seating area so make sure you have easy access for people of all shapes and ages. As you can see, proper designing of this space will result in a truly functional kitchen. Remember, start with the work triangle featuring your appliances and then add the cabinets. Don't forget to add space to store a blender, food processor, toaster, slow cooker, etc. Follow these tried and true kitchen design principles and you will have a kitchen that "just works."

Bathroom Design

Bathrooms are often fit into a design wherever there is space left over. Proper bathroom planning can save you money when you factor in the cost of unnecessary plumbing. A common bathroom design that has been around for decades is what I call the "T design." The bathtub is placed across the width of the room, the five-foot wall. The head of the bathtub, the toilet and the vanity are placed on the long eight-foot wall. This "T" design and forty square foot footprint is economical and still functions well today.

Storage is something that many people forget about when designing a bathroom. Towels, toilet paper, hair dryers, curling irons, and a myriad of personal care products all need to be stored in a bathroom. Vanities will address some storage, along with "over the john" cabinets. If the eight-foot wall opposite the bathroom-plumbing wall is common to a closet, some space can be taken from that closet and a shallow linen closet or open shelves can be recessed into the bathroom wall (preferably behind the door). Medicine cabinets are also useful for storage of medicines and smaller supplies. These cabinets can be recessed or surface-mounted in the front wall or a sidewall. If you choose to mount the cabinet on a sidewall, make sure the door swing is correct so you can use the mirror. There are also double-hinged single doors for added flexibility when in use. If you are going to recess the cabinet, make sure to adjust your plumbing vent pipe and electrical runs to accommodate the recessed cabinet.

A bathroom feature that has been used for storage in the past and has all but disappeared is the soffit above the tub. This unique feature utilizes the space inside the soffit as storage, and is accessible by a door placed in the front of the overhead soffit.

One added note when designing a bathroom is to make sure you correctly size the exhaust fan to eliminate excess moisture. You will find sizing information in the bathroom section. The bathroom exhaust fan should run for twenty minutes after exiting the tub or shower to properly remove excess moisture from the bathroom. A simple solution is to install a timer type switch that will automatically turn off in a predetermined amount of time.

During the electrical design phase of the project, you should think about heating the floor or adding heat lamps in the ceiling to take the chill out of the air when exiting the tub or shower. When you add lighting to the design, be aware that overhead lighting can cause shadows when people are using the mirrors. I suggest you place a light source overhead and also around or at the top of the mirror. Following these basic design tips will help to give you many years of pleasurable use.

Bedroom Design

The bedroom can be far more than a room for a bed and a dresser. You should spend time planning and designing elements in this room. The outcome will make this room comfortable, visually appealing, and multi-functional for many years.

A good place to start in this process is determining what you will use the bedroom for other than sleeping. Will it be used for studying, homework, or watching television? Or is it your private getaway from the kids? Do you need a separate dressing area? Of course you will need to factor in the overall size of the room.

If the bedroom is on the smaller side, there are several designing and decorating techniques you can utilize to make the room appear larger such as:

- Hard-surface flooring to offset bed linens and window treatments
- Mirrored closet doors or a large wall mirror
- A three-dimensional wall mural
- A tray or vaulted ceiling
- Light and simple window treatments

You can also "enlarge" the room by using these tricks:

- Store off-season clothes in another area, such as a guest room or basement
- Install a pedestal under your bed, or low, flat, plastic containers to store folded clothes
- Build a dresser into an attic wall
- Raise the clothes-hanging rod in the closet and put your dresser in the closet
- Install closet organizers
- Utilize twelve-inch deep upper style kitchen cabinets to build a wall of storage cabinets
- Open the ceiling to unused attic space and incorporate a loft
- Bunk beds are a great space-saver and work well for children

The use of lighting helps to highlight and create "different spaces" in your room. Here are a few tips for lighting:

- Install a skylight, either full or the tube type.
- Install track lighting to highlight an area or wall hanging.
- Pendants hanging over your bed will give the room visual depth
- A wall-attached adjustable arm light for reading in bed will also add visual depth and eliminate a table top lamp

The door to the bedroom should be at a minimum of thirty-inches wide to allow furniture to be moved in and out. The optimum placement of the door would have it open against a wall at ninety degrees, rather than have it open one hundred and eighty degrees and take up valuable wall space. Placement of telephone and television jacks should also be thought of in relation to furniture placement.

Spending some time thinking and planning your bedroom is worth the time and effort if you really want it to be a welcoming and comfortable space.

Family Room/Den/Living Room Design

Although I have combined these types of rooms together for the purposes of this book, these rooms are different. The living room and family room are communal or social spaces. The den, study, library, and music room are each more of a singular space and should reflect their purpose as well as the owner's personal taste.

In social or communal spaces, you need to realize that different members of the family will be using the space for different activities. Even if there are only two of you, the space will undoubtedly be used for family gatherings and entertaining. Keeping this fact in mind when you are designing the room will add to its eventual comfort level and functionality.

The number one concern in either of these types of rooms is traffic flow. When walking through or moving from one piece of furniture to another, you should not have to step around another piece of furniture to get where you need to go.

People are always trying to make a room look and feel larger than it is. If your room is the long and narrow type, you can stretch the width by adding rows of boxed beams in a dark or light color. This will add visual stopping points or steps as the eye naturally travels the depth of the room. Here are examples of other tricks or techniques you can use to make the room wider, deeper, or taller:

- Hanging lights such as pendants or track lighting
- Built-in wall units
- Change in wall or ceiling texture
- Floor to ceiling window treatments with vertical stripes add height

If you can help the room become multi-functional to serve more than one purpose, then it will "live larger." One simple way to accomplish this is to use chairs that swivel. The chairs can be facing the television in one setting and turned around to form a conversation area for another type of social gathering. Another multi-functional method is the use of cube-type end tables with flip tops. One side is a cushion that can be used for a seat or footrest and the other side is a hard surface for placing food or drinks. The use of a futon or sleeper sofa is yet another type of multi-functional furniture you can utilize to enlarge the space.

The use of multi-functional furniture also easily allows you to divide the room into separate areas, a technique that is often used when you have a large space and you are trying to have intimate spaces within the larger area. This type of flex grouping will allow different family members or guests to use the space for different purposes.

As with any space in your home when you take the time to plan the space before you begin the physical project, the result will be a comfortable, useful and welcoming space.

Section 2

Building Your Project

Kitchen

Kitchen dimensions assume thirty-four and a half inch tall AFF base cabinets with one and a half inch thick countertop, eighteen inches vertical space between countertop and upper cabinets and minimum thirty inch tall upper cabinets.

Plumbing	Dimension	Comments
Sink drain without disposal	18" AFF to center of drain	
Sink drain with disposal	15" AFF to center of drain	
Water supply	21" AFF and 4" right and left of drain center	
Dishwasher supply	Run from a "T" off of the sink hot water supply with a separate shutoff	Install water line when installing dishwasher
Refrigerator water/ice supply	Center of refrigerator wall opening. Bring water line up within the space (depth) of the base shoe. Up from the floor or side cabinet at rear of refrigerator.	Standard (up to 36") width refrigerators have water connections in the rear. Some larger refrigerators have them located in front. Leave 36" to 48" of coiled line for movement of the refrigerator

Gas Piping	Dimension	Comments
Standard gas stove	Within center 12" of stove opening and no more than 12" AFF. Some newer stoves have a recessed area in the rear of the stove for gas piping to fit into.	Pipe contained within 1 ½" of finished wall: swing valve to the side. Optional; Use a recessed dryer box for gas piping to fit stove tight to wall.
Gas cooktop	Consult manufacturers specifications	
Built-in oven	Consult Manufacturers Specifications	

NOTES

Kitchen continued:

Electrical	Dimension	Comments
GFCI countertop receptacles	Spacing so no point on the countertop is >24" from a receptacle. Countertops ≥ 12" in length require a receptacle. Install GFI 46" AFF to center	Countertops separated by sinks and stoves considered separate countertops. No face-up receptacles are allowed
Definition of peninsula	Long Dimension >24" and short dimension >12" measured from connecting edge	If ≥ 72" and separating rooms must also have a wall height non-GFCI receptacle installed
GFCI island receptacles	Island requires 1 receptacle no 24" rule maximum 6" countertop overhang	No more than 12" below countertop or install in backsplash or cabinet above
GFCI peninsula receptacles	Peninsula requires 1 receptacle no 24" rule	No more than 12" below countertop 6" maximum overhang
Switches above the countertop (between cabinets)	46" AFF to center	Measure from and set cabinets on the same finished floor surface
Light switches not above countertop	51" AFF to center	
Microwave plug in receptacle	6" to center of receptacle measured from bottom of cabinet/top of microwave	10" to center of receptacle right or left from center of microwave. Don't mount in center behind exhaust duct.
Dishwasher connection	32" long Greenfield whip / 1" AFF within center 12" of dishwasher opening	
Electric stove connection	Consult manufacturer specifications	Consult manufacturer specifications
Gas stove electric receptacle 110 volt	20" AFF	4" Right or left of center in the stove space
Refrigerator receptacle	46" AFF	Centered in the refrigerator space

NOTES

Kitchen continued:

Electrical	Dimension	Comments
Garbage disposal connection	Greenfield whip	Rough length to reach bottom center of the sink where disposal mounts
Garbage disposal switch	46" AFF	
Cable/satellite	46" AFF to center	
Under cabinet lighting plug-in receptacle inside of cabinet	62" AFF to center and centered right to left within the cabinet	Switched receptacle located in upper cabinets
Wall-hung telephone above countertop	48" AFF to center	

NOTES

Kitchen continued:

HVAC	Dimension	Comment
Return air	Do not install return air opening in the kitchen	
Microwave/hood fan exhaust	4" flexible or rigid duct vented to the exterior	Consult manufacturers specifications for maximum duct run
Toe kick heat register w/24" deep cabinets w/3" recessed toe kick (standard)	Bring front of duct up through floor no more than 20" (projecting) from finished wall that cabinets will mount to. Do not locate in front of/ under sink or appliances	If necessary cabinet bottom can be framed to act like duct to channel air to the front toe kick/ register

Cabinetry	Dimension	Comments
Standard base cabinets	24" deep/21" deep at toe kick. Toe kick height is 3 ½" to 4 ½" Any dimension is available in custom cabinets. Changing the depth will affect sink and appliances	Box cabinets available at local home improvement stores come in 3" incremental widths generally starting at 12" wide
Lazy Susan corner cabinets	Minimum projection from wall corner is 36"	
Sink base cabinet	Minimum width is 36"	
Dishwasher opening	24" finished opening width	Dishwasher and cabinets to be set on the same finished floor
Stove opening	30" finished opening width	Watch countertop overhang
Refrigerator opening	Minimum 36" finished opening width. If refrigerator is installed next to a wall and the door opens against it, add the depth of the handle and ½ the thickness of the door to the width opening	Refrigerator height will generally vary from 67" minimum to 72"

NOTES

Kitchen continued:

Cabinetry	Dimensions	Comments
Standard upper wall cabinets	12" deep and 30" tall. The depth may increase to 15" or more. Other heights available are 36" and 42" tall.	Cabinets over refrigerators and microwaves can vary in depth.
Cabinet drawer hardware	Handles or knobs should be mounted center height and center width	
Cabinet door hardware	Base cabinet door handles should be centered at 23" AFF and centered in the side rail. The center of the handle should not be more than 4" below the top edge of the door panel.	Upper cabinet door handles should be installed in the center of the door side rail. The center of the handle should not be more than 4" above the bottom edge of the door panel.
Countertops	25" deep from finished wall Back and side splashes are 4" tall	Countertops are typically 1 ½" thick, although this can vary with materials
Pantry shelving	Bottom shelf up to 36" AFF to store dog food/ kitty litter bags Add some 12" apart Some 15" & 20" apart	Adjustable shelving is the best option

Appliances	Dimensions	Comments
Under-the-counter dishwasher	24" wide	Fits under standard height countertop
Gas or Electric-Free Standing Range	30" Wide	
Standard refrigerators	Typically 30" to 36" wide	67" to 72" tall
Under-counter refrigerator	Typically 24" wide	
Wine cooler	Typically 19" to 24" wide	
Trash compactor	12" or 15" wide	Fits under standard countertop. Some brands are plug-in type, others are hard wire type

NOTES

Kitchen continued:

Safety	Specifications	Comments
Fire extinguisher	A rated B rated C rated	A for wood and paper B for grease, gasoline, liquids C for electrical, wires
Fire extinguisher labeled *ABC multi-use*	Should have a multi-use labeled fire extinguisher in the kitchen	
Smoke detector	Ceiling mounted 18" out from wall/corner Wall mounted 4" to 12" below ceiling 18" away from the corner. Mount near kitchen but not in it	Ionization-type detectors near kitchen will give nuisance alarms
Carbon monoxide detector	See manufacturer's specifications	Option would be to mount 1 high and 1 low

NOTES

Bathroom

Vanity Rough In	Height	Horizontal Spacing
Sink wall drain*	18" AFF	Center on sink bowl drain; extra deep bowls can require higher vanity or lower drain
Water supply**	21" AFF	4" from center of sink drain in either direction

* If you have a bank of drawers in your vanity, you may want to offset the rough in to the opposite side of drawers.

**Do not install water supply lines in an exterior wall; come up from the floor.

Toilet /Bidet Rough In	Height/Projection/Depth	Horizontal Spacing
Toilet floor drain	12" from finished back/ tank wall to center	Allow 18" from center to each side for comfortable clearance
Toilet water supply from wall	8" AFF	6" left of center
Toilet water supply from floor	4" from finished wall and 8" AFF	6" left of center
Toilet compartment/room	66" minimum depth of room: with inswing door add width of door to 66"	36" recommended width (30" minimum)
Bidet	See Manufacturer's Specifications	

** Allow minimum 24" clearance from front of the bowl to any object

** If toilet is installed near a door, check door swing against bowl projection

Urinal	Spacing	Comments
Wall mount	15" center to wall or partition	30" center to center. If a partition is added, add the thickness to spacing
Height	See Manufacturer's Specifications	

NOTES

Bathroom continued:

Tub/Shower	Dimensions	Comments
Whirlpools	See Manufacturer	
Tub/shower walls	Waterproof material on the walls shall be at least 72" AFF	
Tub/shower faucet Standard 5' Tubs up to 18" deep	30" AFF Maximum 33" AFF Shower Head 68" to 81" AFF	Height of shower enclosure walls may dictate showerhead height
Tub drain	No standard because of many different style tubs.	Place tub and install drain and overflow piping. Run drain pipe to the tub drain and overflow stub and connect
Shower stall	Minimum shower size 30' x 30" but highly recommended minimum is 36" x 36"	A 30" disk must fit within walls and have no intrusion from seats, etc.
Shower control valve	38" to 48" AFF	
Shower drain	Consult manufacturer	
Shower seat	17" to 19" above shower floor	Seat should be 15" deep

Cabinetry	Dimensions	Comments
Single bowl vanity	Normal is 21" space saver is 18"	24" is minimum, 36" is comfortable width
Double bowl vanity	21" deep	60" minimum
Vanity height (without top)	30" height old standard	34" height is comfortable
Faucet holes	4" o.c. spread	8" o.c. spread are available
Over the toilet cabinet	Cabinet bottom 42" to 48" AFF	

NOTES

Bathroom continued:

Bars and Rings	Dimensions	Comments
Towel bar	48" to 54" AFF Children 36" AFF	Depending on user's height
Towel ring	Bottom of ring 15" above vanity top	
Toilet paper holder	25" AFF to center and 9" to center from front of bowl	
Robe hook	66" AFF	For tall people 70"
Mirror	Center mirror on faucet and center mirror at 54" AFF	The bottom of the mirror reflective surface cannot be more than 40" AFF
Horizontal grab bar in tub or shower	33" to 36" AFF	1 ¼" to 1 ½" diameter and 1 ½" from wall
Horizontal grab bar toilet area side wall	42" long on sidewall / starting no more than 12" off back wall	33" to 36" AFF
Horizontal grab bar toilet area rear wall	36" long, of which a minimum of 24" from center to transfer area (no wall) side	33" to 36" AFF

Glass in Bathrooms	Dimensions	Comments
Windows inside enclosure	Glass must be tempered to a height of 60" AFF and labeled	Glass block may be substituted for tempered glass. Check local code authorities
Shower doors	Must be tempered and labeled	
Windows outside of enclosure	All windows beginning below 18" AFF must be tempered and labeled	Glass block may be substituted for tempered glass. Check local code authorities

NOTES

Bathroom continued:

Electrical	Dimensions	Comments
Receptacles/GFCI outlets	GFCI required within 36" of sink but not within 36" of any tub/shower	Placing a GFCI receptacle over the vanity may require the mirror to be cut
Light switches	All bathrooms must have minimum 1 light switch located by the door (no pull chains)	Not within 36" of tub/shower
Light fixtures	No light fixture including recessed may be installed within 3' horizontally and 8' vertically from the rim of the shower or tub	Most local building officials will permit recessed lighting if it is sealed and waterproof within this space
Light fixture above mirror	Vertical height AFF depends on mirror height and fixture width	Mount fixture center on width of vanity
TV in tub area	No electrical device/fixture can be within 3' horizontally and 8' vertically from the rim of the shower or tub	Swivel mounts work well

NOTES

Bathroom continued:

HVAC	Specifications	Comments
Exhaust Ventilation <100 sq. ft.	1 CFM per 1 sq. ft. of bath area if less than 100 sq. ft.; Minimum 50 CFM	Vented to the outside
Exhaust Ventilation >100 sq. ft.	Add up the separate components in same room to find required CFM Toilet 50 CFM Bathtub 50 CFM Shower 50 CFM Tub/Shower 50 CFM Whirlpool 100 CFM	If any component is in a room separated by a door it will require a separate exhaust fan
Steam shower	Always needs own ventilation	See manufacturer's details
Exhaust ventilation	Run fan for 20 minutes after shower/tub use	Use of a timer is beneficial
Heat	Code Requires 68°	Comfortable is 10° Warmer, when stepping out of tub or shower
Return air	Do not install return air vent in bathroom	

Carpentry	Specifications	Comments
Bathroom door rough opening	2" wider than door slab and 2" taller than door slab	
Door width	There is no minimum but 30" is comfortable	ADA requires a 32" clear opening which requires a 34" door

NOTES

Laundry Room

Items	Dimensions	Comments
Room depth	8" minimum deeper then deepest appliance	Recommended depth is 12" deeper than deepest appliance
Room width	Minimum 6" wider than both units	If the room width is less than the total of 8" plus the unit width plus the unit door width, dryer will have to be offset (or room made larger) to allow the dryer door to open passed the room door
Washer box	Top of washer supply and drain box 48" AFF	For standard top load units
Dryer vent	4" flexible duct	Recessed dryer box is recommended
Gas pipe	Fit into dryer box	Recessed dryer box is recommended
Shelf over top-loading washer	65" AFF	For top-load washers
Floor drain	2" from front edge of appliances to center of drain Centered in the room	Floor drain should be accessible to add water to the drain if necessary because of evaporation
Washer/dryer receptacle	Center of GFCI receptacle 44" AFF	Center GFCI receptacle in the center of the appliances

NOTES

Bedrooms

Rooms/Closets/Windows	Dimensions	Comments
Bedroom minimum	GLA >70 sq. ft. with no single dimension being less than 7' (GLA=gross living area)	If ceiling slopes (think dormer), you can only count floor area that is taller than 7' in the GLA (gross living area)
Bedroom door rough opening	2" wider than door slab and 2" taller than door slab	
Bedroom door	There is no code minimum requirement but 30" is comfortable	Note ADA requires a clear 32" opening which requires a 34" door
Doorstop projecting-type wall mount	1 ½" AFF and 1 ½" inside of door edge	
Bi-fold door handles	36" AFF to center of handle	1 ¼" from center joint into center panel
Wire closet shelving	65" for single shelf	42" and 84" for double shelves
Emergency window egress	5.7 sq. ft. net clear opening minimum size. 24" minimum height and 20" minimum width (clear opening)	Maximum sill height AFF 44"

HVAC

Supply		
GFA supply air register	Centered under window	Supplying heat from a floor register is the best practice
GFA return air register	Opposite side of room from the heat supply	

NOTES

Bedrooms continued:

Electrical	Dimensions	Comments
Switches	51" AFF to center of switch	
Receptacles	15" AFF to center of receptacle	
TV/satellite/cable	15" to center of box	
Ceiling fan	7' AFF to bottom of the blades	
Closet ceiling light Non-walk-in closet	6" forward from center of closet to center of fixture	
Closet ceiling light Walk-in closet	Center the light fixture in the room	
Smoke detector	Ceiling mounted 18" out from wall/corner wall mounted 4" to 12" below ceiling 18" away from corner Mount in every level/ bedroom/near kitchen but not in it	Ionization-type detectors near kitchen will give nuisance alarms
Carbon monoxide detector	See manufacturer's specifications	Option would be to mount 1 high and 1 low

Mattress Type	Dimension	Comments
King mattress size	76" wide and 80" long	Allow for headboard
California king mattress size	72" wide and 84" long	Allow for headboard
Queen mattress size	60" wide and 80" long	Allow for headboard
Full bed size	53" wide and 73" long	Allow for headboard
Twin bed size	38" wide and 75" long	Allow for headboard

NOTES

Family Room/Den/Living Room

Electric/TV/Low Voltage

Device	Dimension	Comments
Wall switches	51" AFF to center of switch	
Receptacle	15" AFF to center of switch	
Phone TV/satellite/cable	15" AFF to center of switch	
Ceiling fan	7' AFF to bottom of blades	
Door bell chime box	6'8" AFF to bottom of box	
Thermostat	56" to center	
Smoke detector	Ceiling mounted 18" out from wall/corner. Wall mounted 4" to 12" below ceiling 18" away from corner. Mount in every level/ bedroom/ near kitchen but not in it	Ionization-type detectors near kitchen will give nuisance alarms
Carbon monoxide detector	See manufacturer's specifications	Option would be to mount 1 high and 1 low

Fireplaces

Fireplaces	Height	
Fireplace hearth	8" to 20" high AFF	Minimum 20" projection from the face of the fireplace
Fireplace mantel	Mantel can project 1/8" out for every 1" of rise (vertical distance) from the firebox	Typical mounting height AFF 48" to 55"
Gas line	Install gas valve 2" above finished hearth height	Consult manufacturer's specifications for fire box entry point

NOTES

Family Room/Den/Living Room continued:

Televisions/Monitors

Screen Size	Viewing Distance Range	Mounting Height
26" Flat Screen TV	3'3" to 6'5" Middle 4'10"	49" to center
30"	3'8" to 7'6" Middle 5' 11"	53" to center
34"	4'3" to 8'5" Middle 6'4"	54" to center
42"	5'3" to 10'5" Middle 7'10"	59" to center
46"	5'8" to 11'5" Middle 8"4"	60" to center
50"	6'3" to 12'5" Middle 9'1"	62" to center
55"	6'8" to 12'8" Middle 9'4"	63" to center
60"	7'5" to 15' Middle 11'3"	66" to center
65"	8'1" to 16'3" Middle 12'1"	69" to center
Or you can use TV viewing distance formula	VD = TVS/.55	VD= Viewing Distance TVS = Television Size
TV mounting height formula	MHC = ELH +(VD *.22)	MHC = Mounting Height Center ELH = Eye Level Height (Measure in viewing chair) VD = Viewing Distance

NOTES

Stairs and Railing

(All dimensions shall be exclusive of carpets, rugs and runners)

Stairs/Railings	Dimensions	Comments
Stairway width	Clear width of 36" above the handrail	Clear width at handrail not less than 31 ½" when handrail is installed on one side. 27" when handrail is installed on both sides
Stairway landing/ platform	Minimum 36" x 36"	Enclosed accessible space under the stairs/platforms to be covered with a minimum ½" drywall
Stairway headroom	Minimum headroom 6'8" measured vertically from a sloped tread line or platform surface	
Tread width	Minimum tread width measured from front of the previous stair nosing to the edge of the tread is 10"	All tread widths must be within 3/8" of each other
Thread nosing	Required nosing on all solid riser stairs to be not more than 1 ¼" or less than ¾". The greatest nosing projection to the least shall not exceed 3/8"	Where treads are minimum of 11" no nosing is required
Riser height	Maximum riser height shall be 7 ¾"	All riser heights must be within 3/8" of each other
Open risers	Open risers are permitted if a 4" diameter sphere cannot pass through opening created by open riser	

NOTES

Stairs and Railing continued:

Stairs/Railings	Dimensions	Comments
Handrail	Must be provided where there are four or more risers	Handrails Mounted To A wall must have minimum 1 ½" between wall and handrail/no projection more than 4 ½"
Handrail height	Measured vertically from a sloped tread line not less than 34" or more than 38"	Handrail must be continuous for entire run
Handrail grip	If circular, no less than 1 ¼" or more than 2"	
Handrail grip	Perimeter dimension minimum of 4" and maximum of 6 ¼" Maximum cross section gripable dimension of 2 ¼"	Minimum radius on corners of .01
Handrail start/end points	Beginning at edge of top tread (first riser) and ending above last thread edge	Handrail must be returned to wall or ended at a newel post
Guard/railing	Required if the difference between the grade and platform is more than 30" at any point within 36"	Railing height shall not be less than 34" measured vertically from a sloped thread line. If top of rail is used for a handrail, height shall not exceed more than 38" from same line
Guard/railing spacing	Vertical members of required guard shall be spaced so a 4" diameter sphere shall not pass through openings	Exception: triangular opening formed by the tread, riser, and bottom of guard shall not allow a 6" diameter sphere to pass through

NOTES

Stairs and Railing continued:

How to. . .

Calculate The Riser And Run For A Staircase		
Step 1. Find the total vertical rise in inches (finish floor to finish floor)	Step 2. To find number of risers needed (divide total vertical rise by 7")	Step 2a. Now take the answer from #2 and round up or down to the nearest whole number (14, 15, 16, etc.). This is total risers needed
Step 3. To find the rise of every step, divide the total rise (in inches) by the whole number of risers that was the answer of #2a.	Step 4. Find number of treads required. It is always one less than the number of risers	Step 5. To find the total run of staircase, multiply the number of treads by the ideal tread width
Step 6. Adjust run/tread width as needed to meet your floor plan needs, but remember to stay within the building code's guidelines		

To watch my "How to Layout and Cut a Stair Stringer" Video visit ContractorJohn.com, select "My Blog" and click on "How-To Videos".

NOTES

Section 3

Remodeling Mathematics

EZ Math Formulas and Tables

Area/Length	Formula	Comments
 H ▲ △ ▼ Base	Area= base x height/2	For gable end area For hip roof surfaces area
A B C	Length of C = $a^2 + b^2 = c^2$ Area = a*b/2	For the slope length of a roof plane
A B	Area = a * b	For the area of any rectangle or square
Height ▲ ▼ Base	Area = base * height	Area of any plane where one side is equal to another side and the other two sides are equal to each other
B Height D	Area=(b +d)*h/2	Area of a mansard roof side

NOTES

EZ Math Formulas and Tables continued:

Volume/Circumference Formula	Formula	Comments
Radius ⟶ Height	Volume = ∏*r2*h Circumference = 2 *(∏*r)	(∏ = 3.14) Results = Cubic units 1 cubic yard = 46,656 cu. inches 1 cubic foot = 1,728 cu. inches
	Volume = length*depth*height	Results = cubic units 1 cubic yard = 46,656 cu. inches 1 cubic foot = 1,725 cu. inches
Height Base	Volume = 1/3 * (b*h)	Results = cubic units

NOTES

EZ Math Formulas and Tables continued:

Roofing

(L*W)* Multiplier = Base Squares + Cap & Waste Factor= Total Squares Needed
Shingles (1 square of shingles will cover 100 sq. ft. of roof area)

Formula To Find Roof Squares	Multiplier Pitch	Cap and Waste Factor
Length * Width (include overhang) of building	1.03 for 3/12 pitch	Gable add 10% cap & waste. Hip add 17% cap & waste
Length * Width (include overhang) of building	1.05 for 4/12 pitch	Gable add 10% cap & waste. Hip add 17% cap & waste
Length * Width (include overhang) of building	1.08 for 5/12 pitch	Gable add 10% cap & waste. Hip add 17% cap & waste
Length * Width (include overhang) of building	1.12 for 6/12 pitch	Gable add 10% cap & waste. Hip add 17% cap & waste
Length * Width (include overhang) of building	1.21 for 8/12 pitch	Gable add 10% cap & waste. Hip add 17% cap & waste
Length * Width (include overhang) of building	1.25 for 9/12 pitch	Gable add 10% cap & waste. Hip add 17% cap & waste
Length * Width (include overhang) of building	1.30 for 10/12 pitch	Gable add 10% cap & waste. Hip add 17% cap & waste
Length * Width (include overhang) of building	1.42 for 12/12 pitch	Gable add 10% cap & waste. Hip add 17% cap & waste
Length * Width (include overhang) of building	1.81 for 18/12 pitch	Gable add 10% cap & waste. Hip add 17% cap & waste

NOTES

EZ Math Formulas and Tables continued:

Roofing Paper/Roofing Felt

Type of Material	Coverage	Comments
15# roofing felt	1 roll per 4 sq. of shingles	Overlap 6", use guidelines on felt and overlap 12" on butt ends
30# roofing felt	1 roll per 2 sq. of shingles	
Button cap nails	Approximately 30 per roll of felt	3 to 4 button cap nails on the edge then space 3' apart top and bottom and 1 in center of 3' to form an "X" pattern

Attic Ventilation

Most building departments/codes use 1/150 rule 60% of which is for intake and 40% for exhaust. (1 sq. ft. of "net free area" (NFA) per 150 sq. ft. of floor attic space) (NFA is affected by mesh size)

Attic Floor Space Area sq. Ft.	Total Ventilation Needed (NFA)	12"x12" Eve Vents (= .8 sq. ft. NFA) Needed 1/8" Mesh Screen** Intake 60% of Total NFA	12"x12" Standard Roof Vents (= .7 sq. Ft. NFA) Needed Exhaust 40% of Total NFA
1000 sq. ft.	7 sq. ft.	5*	4
1200 sq. ft.	8 sq. ft.	6*	5
1400 sq. ft.	10 sq. ft.	8*	6
1600 sq. ft.	11 sq. ft.	8*	6
1800 sq. ft.	12 sq. ft.	9*	7
2000 sq. ft.	14 sq. ft.	10*	8
2200 sq. ft.	15 sq. ft.	11*	8
2400 sq. ft.	16 sq. ft.	12*	9
2600 sq. ft.	18 sq. ft.	13*	10

*All Numbers are rounded up

** Number of eve vents was calculated using vents with 1/8" mesh screen.

NOTES

EZ Math Formulas and Tables continued:

HVAC Formulas and Tables

The number of windows, insulation, sunlight, etc., can and will influence the general guidelines below. The efficiency of the furnace is also something to include in your calculations. A 100,000 BTU furnace input rating on an 85% efficient furnace will have an output BTU rating of 85,000 BTU's. The actual output rating is the correct number to use when calculating size.

Input BTU rating * Efficiency = BTU Output

HVAC Formulas/Tables	Calculation	Comments
Calculate CFM	L xW x H	Approximately 1 CFM of air is needed per 1 sq. ft. of area
Calculate heating BTU's	30 to 40 BTU's per CFM in warmer climates	45 to 60 BTU's per CFM in colder climates
Calculate cooling BTU's	1 Ton = 12,000 BTU's	12,000 BTU's is the amount or power needed to melt one ton of ice in one hour
Area To Be Cooled	BTU's Needed	Tonnage Needed
100 to 150 sq. ft.	5,000 BTU's	
150 to 250 sq. ft.	6,000 BTU's	
250 to 300 sq. ft.	7,000 BTU's	
300 to 350 sq. ft.	8,000 BTU's	
350 to 400 sq. ft.	9,000 BTU's	
400 to 450 sq. ft.	10,000 BTU's	
450 to 550 sq. ft.	12,000 BTU's	1 Ton
550 to 700 sq. ft.	14,000 BTU's	
700 to 1000 sq. ft.	18,000 BTU's	1 ½ Ton
1000 to 1200 sq. ft.	21,000 BTU's	
1200 to 1400 sq. ft.	23,000 BTU's	
1400 to 1500 sq. ft.	24,000 BTU's	2 Tons
1500 to 2000 sq. ft.	30,000 BTU's	
2000 to 2500 sq. ft.	34,000 BTU's	

If using a window a/c unit and the room you are cooling is regularly used by more than two people, add 600 BTU's for each person over two.

If you are cooling a kitchen, add 4,000 BTU's to the number above

NOTES

EZ Math Formulas and Tables continued:

SEER ratings of a unit refer to the energy efficiency of the unit. The higher the number, the more efficient it is, meaning it uses less electricity to run. As of 2006, the minimum allowable rating is 13, although that will jump to a 14 SEER January 1st, 2015. Central air conditioning units are currently available in a 20.5 SEER. Currently window units are exempt from this requirement and remain around 10 SEER.

CFM Table (rates may vary by manufacturer and material)

CFM	Round Duct	Rectangle Duct	Supply Register	Return Grille
60	5	2 ¼ x 10	4 x 10	N/a
100	6	2 ¼ x 12	4 x 10	6 x 10
150	7	3 ¼ x 14	4 x 14	6 x 12
200	8	4 x 14	6 x 14	6 x 14
300	9	8 x 8	8 x 14	8 x 14
400	10	8 x 10	N/A	8 x 18
500	12	8 x 12	N/A	8 x 24
600	12	8 x 14	N/A	12 x 18
700	14	8 x 16	N/A	14 x 18
800	14	8 x 18	N/A	10 x 30
900	14	8 x 20	N/A	14 x 24
1000	16	8 x 22	N/A	12 x 30
1200	16	8 x 24	N/A	18 x 24
1400	16	8 x 28	N/A	18 x 30
1600	18	8 x 32	N/A	20 x 30
1800	20	10 x 28	N/A	24 x 30
2000	20	10 x 30	N/A	24 x 30

NOTES

EZ Math Formulas and Tables continued:

Drainage Fixture Unit (d.f.u.)

***Any drainpipe smaller than 4" must have a ¼" per foot slope

Type of Fixture or Group of Fixtures	DFU Value
Bar sink	1
Bathtub (with or without shower head and/or whirlpool)	2
Bidet	1
Clothes washer standpipe	2
Dishwasher	2
Floor Drain a	0
Kitchen sink	2
Lavatory	1
Laundry tub	2
Shower stall	2
Water closet/toilet (1.6 gallons per flush)	3
Water closet/toilet (greater than 1.6 gallons per flush)	4
Full bathroom group with 1.6 gallon water closet and bathtub with or without shower head and whirlpool	5
Full bathroom group with greater than 1.6 gallon water closet and bathtub with/without showerhead and whirlpool	6
Half bath group (1.6 gallon water closet and lavatory)	5
Half bath group (greater than 1.6 gallon water closet and lavatory)	6
Kitchen group (kitchen sink with or without garbage disposal and dishwasher)	2
Laundry group (clothes washer standpipe and laundry tub)	3
Multiple group bathrooms b 1.5 Baths 2 Baths 2.5 Baths 3.0 Baths 3.5 Baths	 7 8 9 10 11

A = A floor drain in itself adds no load, except when used as a receptor, then use appropriate fixture value

B = Add 2 d.f.u. for each additional full bath

NOTES

EZ Math Formulas and Tables continued:

Maximum fixture units to be connected to branches or stacks

Nominal Pipe Size Inches	Any Horizontal Fixture Branch	Any One Vertical Stack or Drain
1 ¼" a	---	---
1 ½" b	3	4
2" b	6	10
2 ½" b	12	20
3"	20	48
4"	160	240

A = 1 ¼" pipe limited to a single fixture drain or trap arm

B = no water closets

Horizontal piping shall slope no less than ¼" (2%) per 12" of run for pipes 2 ½" in diameter and less and not less than 1/8" (1%) per 12" of run for pipes with diameters 3" and larger.

NOTES

EZ Math Formulas and Tables continued:

Electrical Wire Uses and Sizing (per 50' run)

For copper wire only. Do not use with aluminum wire.

**For every additional 50 feet of cable or wire run up to #8 you should upgrade to the next size. **Add the Length of Any Extension Cords To Run Length

**The stated safe capacity that the National Electric Code (NEC) recommends at 100% is actually 80% of periodic maximum load.

Gauge or Wire Size	Rated Amperage	Common Uses
18	10	Low-voltage lights/cords
16	13	Light extension cords
14	15	Lighting circuits and fixtures/general use outlets
12	20	Kitchen and bathroom outlets, sump pumps, appliances, and 110 volt room A/C units
10	30	Electric clothes dryers, water heaters, built-in ovens and 220 volt A/C
8	45	Electric cook tops
6	60	Small electric furnaces
4	80	Large electric furnaces and water heaters
2	100	Electrical sub panels
1/0	150	Electrical service panels
2/0	200	Electrical service panels

Length of Wire Run/Wire Size

Amps	100'	150'	150'-200'	200'-250'	250'	250'-300'	300'	300'-400'	400'	400'-500'	500'
12	#12	#10	-	#8	-	-	#6	-	-	#4	-
16	#10	#8	-	#6	-	-	-	#4	-	#2	-
24	#8	-	#6	-	-	#4	-	-	-	#2	-
32	#6	-	#4	-	-	#2	-	-	#1	#1/0	-
40	#6	-	#4	-	#2	-	#1	-	#1/0	-	#2/0

NOTES

EZ Math Formulas and Tables continued:

Electric Formulas

To Find	Formula	A Common Use
Amps	Watts/Volts = Amps	To determine how many light fixtures or outlets you can put on a single circuit (run)
Watts	Volts*Amps = Watts	Can be used to help calculate load

NOTES

Conversion Methods

Fractions

Fractions	1/8	1/4	3/8	1/2	5/8	3/4	7/8	8/8
Decimal	.125	.25	.375	.50	.625	.75	.875	1.0

Inches

Inches	1"	2"	3"	4"	5"	6"	7"	8"	9"	10"	11"	12"
Decimal	.08	.17	.25	.33	.42	.50	.58	.67	.75	.83	.92	1.0

Cubic Units

Unit	Conversion	Formula
Cubic yard =	46,656 cubic inches	H*W*L ((36*36(*36)
Cubic yard =	27 cubic feet	H*W*L ((3*3)*3)
Cubic foot =	1,728 cubic inches	H*W*L ((12*12)*12)

Concrete Conversions

Size Bag of Concrete Mix	Yield in Cubic Feet	Comments
60 lb. bag	.45 cubic foot	Does not factor in waste or irregularity in container or space
80 lb. bag	.6 cubic foot	Does not factor in waste or irregularity in container or space

Liquid Conversions

Unit	Ounces	Comments
Gallon	128	1 Bushel = 8 Gallons
½ Gallon	64	
Quart	32	
Pint	16	
Cup	8	
2 Tablespoons	1	
6 Teaspoons	1	
Liter	------------------------	0.0264 Gallons

NOTES

Conversion Methods continued:

Dilution Ratios

(Fractions of ounces are rounded up to the next whole number)

Ratio	Amount of Additive To 1 U.S. Gallon
1:256	½ of an ounce per gallon
1:200	2/3 of an ounce per gallon
1:128	1 ounce per gallon
1:64	2 ounces per gallon
1:40	3 ounces per gallon
1:30	4 ounces per gallon
1:26	5 ounce per gallon
1:20	6 ounces per gallon
1:16	8 ounces per gallon
1:12	11 ounces per gallon
1:10	13 ounces per gallon
1:8	16 ounces per gallon
1:4	32 ounces per gallon

Coverage for Mulch/Compost, etc.

1 Cubic Yard Will Cover	
324 sq. Ft.	1" thick
162 sq. Ft.	2" thick
108 sq. Ft.	3" thick
81 sq. Ft.	4" thick
65 sq. Ft.	5" thick
54 sq. Ft.	6" thick

A Single 3.0 Cubic Foot Bag Will Cover	
36 sq. Ft.	1" thick
18 sq. Ft.	2" thick
12 sq. Ft.	3" thick
9 sq. Ft.	4" thick
7 sq. Ft.	5" thick
6 sq. Ft.	6" thick

NOTES

Conversion Methods continued:

Convert Common Materials from Cubic Yards to Tons

Material	Cubic Yard	Weight in Tons
Topsoil	1	1.0
Mulch	1	1.0
Compost	1	.5
Gravel	1	1.4
Limestone	1	1.4
Decorative Stone	1	1.3
Dirt	1	1.3
Sand	1	1.5

If the material is wet the net yield will be less, and stone absorbs water.

Common Land Measurements

Parcel	Area	Comments
1 Acre	43,560 sq. ft.	Approximately 208.5' x 208.5'
640 Acres	1 sq. Mile	
1 Section	1 sq. Mile	
1 Township	36 sq. Miles	
5,280 ft.	1 Mile	
1 Rod	11 cubits or 16.5 ft.	Same length as a perch or pole
1 Chain	4 Rods	

Firewood Measurements

Unit Name	Size In Cubic Feet	Conventional Size
Full or bush cord	128 cubic feet	8' long x 4' wide x 4'high
1/2 cord	64 cubic feet	8' long x 2' wide x 4' high
¼ cord	32 cubic feet	4' long x 2' wide x 4' high
Face or rick cord	Not legally defined	Typically but not always 8' long x 4' high x 16" thick

NOTES

Conversion Methods continued:

Nail Sizes and Count per Pound

*Approximate

Nail Size	Length In Inches	*Common Nails Per lb.	*Box Nails Per lb.	*Finish Nails Per lb.
2d	1"	876	1010	1351
3d	1 ¼"	568	635	807
4d	1 ½"	316	437	548
6d	2"	181	236	309
8d	2 ½"	106	145	189
10d	3"	69	94	121
12d	3 ¼"	64	87	113
16d	3 ½"	49	71	90
20d	4"	31	52	62
30d	4 ½"	20	---------	---------

For more detailed information about nail types and uses, visit contractorjohn.com select the "Blog" and search "Nails".

NOTES

Floor/Ceiling Joist Span Table

Floor Joists: 10 PSF Dead Load/40 PSF Live Load

(PSF = pounds per square foot) Dead load is the weight of the actual structure itself. Live load are the items that can be moved into the space, people, furniture, etc.

Lumber		SPF		SYP		Hem-Fir		Doug-Fir	
Size Inches	Spacing Inches	No. 1	No. 2	No. 1	No. 2	No. 1	No. 2	No. 1	No. 2
2 x 6	12	10-6	10-3	10-11	10-9	10-6	10-0	10-11	10-9
	16	9-6	9-4	9-11	9-9	9-6	9-1	9-11	9-9
	24	8-4	8-1	8-8	8-6	8-4	7-11	8-8	8-1
2 x 8	12	13-10	13-6	14-5	14-2	13-10	13-2	14-5	14-2
	16	12-7	12-3	13-1	12-10	12-7	12-0	13-1	12-7
	24	11-10	10-3	11-5	11-10	10-9	10-2	11-0	10-3
2 x 10	12	17-8	17-3	18-5	18-0	17-8	16-10	18-5	17-9
	16	16-0	15-5	16-9	16-1	16-0	15-2	16-5	15-5
	24	14-0	12-7	14-7	13-2	13-1	12-5	13-5	12-7
2 x 12	12	21-6	20-7	22-5	21-9	21-6	20-4	22-0	20-7
	16	19-6	17-10	20-4	18-10	18-7	17-7	19-1	17-10
	24	17-0	14-7	17-5	15-4	15-2	14-4	15-7	14-7

SPF = Spruce Pine Fir
SYP = Southern Yellow Pine
Hem-Fir = Hemlock Fir
Doug-Fir = Douglas Fir

Ceiling Joists: 10 PSF Dead Load/20 PSF Live Load (drywall ceiling/limited attic storage)

Lumber		SPF		SYP		Hem-Fir		Doug-Fir	
Size Inches	Spacing Inches	No. 1	No. 2	No. 1	No. 2	No. 1	No. 2	No. 1	No. 2
2 x 4	16	8-9	8-7	9-1	8-11	8-9	8-4	9-1	8-9
	24	7-8	7-2	8-0	7-8	7-6	7-1	7-8	7-2
2 x 6	16	13-9	12-10	14-4	13-6	13-5	12-8	13-9	12-10
	24	12-0	10-6	12-6	11-0	10-11	10-4	11-2	10-6
2 x 8	16	18-2	16-3	18-11	17-5	16-11	16-0	17-5	16-3
	24	15-10	13-3	15-11	14-2	13-10	13-1	14-2	13-3
2 x 10	16	23-2	19-10	23-2	20-9	20-8	19-7	21-3	19-10
	24	19-5	16-3	18-11	17-0	16-11	16-0	17-4	16-3

NOTES

Header Spans

All headers constructed with Douglas Fir-Larch, Hem-Fir, Southern Pine, and Spruce Pine Fir and required jack studs.

Headers Supporting Roof and Ceiling w/30 lb. ground snow load up to 28' Building Width

Header Size	Span	Jack Studs Required (each side)
2- 2x8	5'11"	2
2- 2x10	7' 3"	2
2- 2x12	8'5"	2
3- 2x8	7'5"	1
3- 2x10	9'1"	2
3- 2x12	10'7"	2

Headers Supporting Roof, Ceiling, and One Center Bearing Floor w/30 lb. ground snow load up to 28'Building Width

Header Size	Span	Jack Studs Required (each side)
2- 2x8	5'0"	2
2- 2x10	6'2"	2
2- 2x12	7'1"	2
3- 2x8	6'3"	2
3- 2x10	7'8"	2
3- 2x12	8'11"	2

Headers Supporting Roof and Ceiling and Two Center Bearing Floors w/30 lb. ground snow load up to 28' Building Width

Header Size	Span	Jack Studs Required (each side)
2- 2x8	4'2"	2
2- 2x10	5'1"	2
2- 2x12	5'10"	3
3- 2x8	5'2"	2
3- 2x10	6'4"	2
3- 2x12	7'4"	2

NOTES

Header Spans continued:

Headers Supporting Roof, Ceiling w/30 lb. ground snow load up to 36' Building Width

Header Size	Span	Jack Studs Required (each side)
2- 2x8	5'4"	2
2- 2x10	6'6"	2
2- 2x12	7'6"	2
3- 2x8	6'8"	1
3- 2x10	8'2"	2
3- 2x12	9'5"	2

Headers Supporting Roof, Ceiling and One Center Bearing Floor w/30 lb. ground snow load up to 36' Building Width

Header Size	Span	Jack Studs Required (each side)
2- 2x8	4'6"	2
2- 2x10	5'6"	2
2- 2x12	6'5"	2
3- 2x8	5'8"	2
3- 2x10	6'11"	2
3- 2x12	8'0"	2

Headers Supporting Roof, Ceiling and Two Center Bearing Floors w/30 lb. ground snow load up to 36' Building Width

Header Size	Span	Jack Studs Required (each side)
2- 2x8	3'9"	2
2- 2x10	4'7"	2
2- 2x12	5'3"	3
3- 2x8	4'8"	2
3- 2x10	5'8"	2
3- 2x12	6'7"	2

NOTES

Rafter Spans

Rafter Spans: 15 PSF Dead Load/30 PSF Live Load (drywall ceilings/all slopes)

Lumber		SPF		SYP		Hem-Fir		Doug-Fir	
Size Inches	Spacing Inches	No. 1	No. 2	No. 1	No. 2	No. 1	No. 2	No. 1	No. 2
2 x 6	16	12-0	11-3	12-6	11-9	11-9	11-1	12-0	11-3
	24	10-6	9-2	10-11	9-7	9-7	9-1	9-10	9-2
2 x 8	16	15-10	14-3	16-6	15-3	14-10	14-0	15-3	14-3
	24	13-10	11-8	13-11	12-5	12-1	11-6	12-5	11-8
2 x 10	16	20-2	17-5	20-3	18-2	18-1	17-2	18-7	17-5
	24	17-0	14-2	16-6	14-10	14-10	14-0	15-2	14-2
2 x 12	16	24-1	20-2	24-1	21-4	21-0	19-11	21-7	20-2
	24	19-8	16-6	19-8	17-5	17-2	16-3	17-7	16-6

SPF = Spruce Pine Fir
SYP = Southern Yellow Pine
Hem-Fir = Hemlock Fir
Doug-Fir = Douglas fir

See individual manufacturers for roof truss ratings and spans.

NOTES

Steel Span Chart

Commonly Used Steel Beams for Residential Basements

Steel Span Chart	1500#/lin. Ft. Load 2 floors max Unless Noted	Support at Ends or Mid Point
W8 x 18	12'0"	Foundation wall or 3 ½" concrete filled steel column
W8 x 21	13'6"	Foundation wall or 3 ½" concrete filled steel column
W8 x 24	15'0"	Foundation wall or 3 ½" concrete filled steel column
W8 x 31	16'0"	Foundation wall or 3 ½" concrete filled steel column
W10 x 22	15'0"	Foundation wall or 3 ½" concrete filled steel column
W10 x 26	16'6"	Foundation wall or 3 ½" concrete filled steel column
W12 x 26**	19'6"**	Foundation wall or 3 ½" concrete filled steel column
W12 x 30**	22'0"**	Foundation wall or 3 ½" concrete filled steel column
W14 x 34**	25'0"**	Foundation wall or 3 ½" concrete filled steel column
W16 x 40**	29'0"**	Foundation wall or 3 ½" concrete filled steel column

** These beams figured at 1000#/lin. Ft. because they are used for garage beams and only support one floor and/or a roof.

NOTES

Section 4

Exterior of the House

Sidewalks and Driveways

Sidewalks

Sidewalk Type	Width Options	Slope
Service walk/path	24" wide	
Sidewalk	36" wide allows for comfortable walking. 48" wide allows for comfortable side by side walking	Maximum comfortable slope is 10%
ADA compliant side walk	60" wide allows two wheel chairs to pass in opposite directions	Maximum slope allowed is 8.33% Cross slope limited to 2%
ADA ramp	36" minimum "clear space width" and a 60"x60" flat platform rest area every 30'	1" rise for every 12' of run 8.33% include a 60"x60" flat platform rest area every 30'

Driveways

Driveway	Width Options	Slope
Single lane	10' to 12' wide, if walls, planters, etc. are on the sides of the driveway width should be increased to 14'	Comfortable grade slope is 5%, maximum is 10%. Acceptable cross slope is 3%. ADA cross slope cannot exceed 2%
Double lane	20' to 24' wide if walls, planters, etc. are on the sides of the driveway width should be increased to 24' to 28'	Comfortable grade slope is 5%, maximum is 10%. Acceptable cross slope Is 3%. ADA cross slope cannot exceed 2%
Parking stall	10' wide x 20' long	
Turn around space	10' wide x 20' long	

NOTES

Grading and Drainage

Grading

Finished Grading	Initial Slope	Secondary Slope
Finished Grade, around your home, including landscaping, should be 4" below TOF	Grade/slope away from your home should be 5% or 6" in the first 10'	After the first 10' slope should be 2% or ¼" per foot

Drainage

Item Discharging	Specification	Comments
Sump pump discharge	1 ½" discharge pipe from the house transitions into a 4" pipe which ends no less than 5' from the house foundation	Transition must have a vertical physical air gap of no less than 2"
Gutters	Downspout should transition into a 4" pipe which ends no less than 5' from the house foundation	Downspouts may be directly connected to, and transition to the 4" drainage pipe
Any solid surface such as sidewalks/patio/stoop	Must be sloped/cross slopped or pitched away from the foundation a minimum of 1/8" per foot	

NOTES

Exterior Bonus Material

Viewing Distance Chart for Address Letters and Exterior Signage
  These distances are generally accurate. Font, color, spacing, etc. can affect ultimate viewing distance.

Letter Height	Best Viewed Distance	Maximum Viewed Distance	Amount of Letters per Horizontal Foot of Sign Space
3"	30'	100'	4 ½
4"	40'	140'	4
5"	50'	180'	3
6"	60'	200'	2 3/4
7"	70'	220'	2 ½
8"	80'	320'	2
9"	90'	360'	1 ¾
10"	100'	425'	1 ¾
12"	120'	500'	1

Multiply letter height * 10 = Best viewing height in feet

Door Bells

Unit	Height	
Door bell button	44" to 48" above finished grade	ADA maximum height is 48"
Door bell chime unit	6'8" to bottom of the chime box	

NOTES

Exterior Bonus Material continued:

Exterior Light Fixtures

Exterior fixtures come in a wide variety of sizes such as hanging and up-mounted. It is advisable to purchase fixture(s) before rough electric work is completed.

Type	Height	Comments
Entry way fixtures	67" to 72" center of fixture above finished grade	6" right of left of the door frame
Security lights	8' to 9' above finished grade	Rule of thumb: Measure the height of the mounted fixture and multiply it by 4. The answer is how far the light should be away from the property line.

Mailboxes

Mailbox Type	Height	Setback
Post mounted by road	41" to 45" of finished grade	Front lip of mailbox to be 6" to 8" back from the road side of the curb or edge of roadway
House mounted	Bottom of mailbox to be 48" off finished grade	

NOTES

Acknowledgements

Dropping a stone into a pond begins the rippling effect, an effect we don't always see the results of. . . and so it is with the people we meet in our life. Some have a greater effect on us, but everyone we meet has some effect on us.

"We are the sum, the collective total of all our life experiences; the good, the bad, and the ugly." -John A. Knoelk

Obviously a book like this is never the work of one person. There are countless people who I have met along this journey of life who have influenced me. Some of these people I would like to specifically acknowledge:

My wife Sue, who not only encouraged me to finish this project, she has also given me the time and whatever resources I needed; and for those thought provoking questions, that when you get to the answer, you know you are a better person for it. I love you.

My mom and dad who taught me hard work is not something to be afraid of. Also, your word means something.

Pastor Paul Schwartz and Pastor Andy Combs whose words would always cause me to dig deeper into my faith.

Nancy Sayer, you will never know how much you helped me to grow in so many areas of my life.

To Sharon Knoelk, who was by my side when it all started, working weekends and holidays, whatever it took.

My son Justin who God is using to teach me, that he is God's son first, then mine, which the knowledge and acceptance of can only strengthen my faith.

To the countless people I have worked with and for.

To all my teachers and mentors, a profound thank you for all the learning opportunities you have put in front of me.

To all my business partners and colleagues who saw enough in this high school graduate, to not only let me run with the ball, they let me use it to start my own game.

"Never discard any meeting or encounter as inconsequential, they are always a part of God's plan." -John A. Knoelk

Additional Resources

www.NAHB.org

The National Association of Homebuilders Website that is full of resources for builders, and remodelers.

www.Remodeling.HW.net

Remodeling Magazine Annual Cost vs. Value Report. There isn't another resource that comes close to the region/city/project-specific information you will find here.

www.NARI.org

The National Association of the Remodeling Industry website that has a whole area devoted to homeowners.

ContractorJohn.com

The home of Contractor John. Here you will find an up-to-date list of additional resources for the remodeler and DIY'er.

DimensionBible.com

Visit the Dimension Bible for updates to the book, and links to other resources

Houzz.com

Called the "Wikipedia of interior and exterior design" by CNN, Houzz provides people with everything they need to improve their homes from start to finish.

Visit ContractorJohn.com for an expanded and ever-changing list of online resources for the remodeling professional, homeowner and DIY'er.

Final Thought

There was a man who was an efficient builder. He had worked for years in a large company and had reached the age of retirement. His employer asked him to build one more house; this was to be his last commission.

The builder took the job, but his heart was not involved. He used inferior materials, the timber was poor, and he failed to see the many things that should have been clear to him had he shown even his normal interest in the work.

When the house was eventually finished, his employer came to him and said, "The house is yours, here is the key, it is a present from me!" The builder immediately regretted that he had not used the best materials and engaged the most capable workers. If only he had known the house was for him. . .

We as God's children are building our own spiritual houses. The faithful God provides us with the best materials. What do we use? Let us be mindful of the many gifts of grace of our Lord. Let us ensure that our hearts participate in everything we do. Not one of us builds a house in eternity for someone else; it is always for ourselves.

Author Unknown